DAMN YOU, AUTOCORRECT!

Messages	**Bff**	Edit

Docking autocorrect

Fucking sickos core

God what the duck

Mar 8, 2013, 9:00 PM

oh my ducking gig

Oh my diving god

Oh my fucking god what the hell is edit with suck its?!

JESUS CHRIST!

DAMN YOU, AUTOCORRECT!

More hilarious text messages you didn't mean to send

Lyndsey Saul

Virgin BOOKS

4 6 8 10 9 7 5 3

First published in the UK in 2013 by Virgin Books,
an imprint of Ebury Publishing
A Random House Group Company

Addresses for companies within The Random House Group Limited
can be found at www.randomhouse.co.uk/offices.htm

The Random House Group Limited Reg. No. 954009

A CIP catalogue record for this book is available from the British Library
The Random House Group Limited supports the Forest Stewardship
Council® (FSC®), the leading international forest-certification organisation.
Our books carrying the FSC label are printed on FSC®-certified paper.
FSC is the only forest-certification scheme supported by the leading
environmental organisations, including Greenpeace. Our paper
procurement policy can be found at www.randomhouse.co.uk/environment

Printed and bound in Great Britain by CPI Group (UK) Ltd,
Croydon, CR0 4YY

ISBN: 9780753541999

Contents

Introduction

DAMN YOU AUTOCORRECT is back in book form, ready to wow you with all the amazing, confusing, strange and side-splitting wonders that autocorrect still manages to produce no matter how hard we try to fight against it.

Since taking over DYAC, I've read thousands and thousands of submissions. Some are common autocorrects we've seen many times (e.g. "animal" or "a nap" to "anal") and some are entirely new combinations of words that literally have me laughing out loud (I won't ruin them by giving you examples here). This is one of the very best things about my job – just when I think I've read all the autocorrects there ever will be, I'm blindsided with something entirely original that I can't wait to share. Rocket tickler, anyone?

MY LIFE WITH AUTOCORRECT

I'm an editor by nature and by profession, so even before I began running Damn You Autocorrect I was an avid backspacer, correcting all my unwanted autocorrects before sending out a text. So, it's only natural that I would become that much more anal (not autocorrect) when my job became reading the humiliating hilarious mistakes submitted by you dear readers, right?

Actually, it's the opposite. I've let my guard down in hopes that something I type will autocorrect into something amazing. In fact, I think I may have developed a yet-to-be diagnosed disorder I call "autocorrect envy". It's not that I've never been autocorrected. It's just that the stars haven't aligned for me to be the unintentional author of one very, very funny autocorrect. I continue to have faith that I one day will. And I refuse to make one up in the meantime!

Which brings me to …

THEY'RE REAL, AND THEY'RE SPECTACULAR

For every one person who loves an autocorrect, there is always another who will call it a fake. "That's fake! Autocorrect never changes _____ to _____ on my phone!" Well, guess what? You're right, and it probably never will for a number of reasons, but primarily because you are on your phone. What many people don't know (even the most prolific of technophiles) is that though every phone's dictionary is created equal when fresh out of its box, once you start using it, all bets are off. Your dictionary learns your colloquialisms as you type them; your slang becomes its slang. Your misspellings, abbreviations and phrasesyoutypewithoutspaces or put in ALL CAPS are remembered by it. Later, your habits come back to haunt you (don't they always?) when autocorrect is just trying to do its job … of making you look stupid.

Autocorrect non-believers must also take human limitations into account. Your thumbs may be narrower than your fellow humans, or you may just be a more adept typist. Along with the physical challenges others face, spectacular autocorrects can also occur when a person isn't hitting the keys they intend to; often they are spelling a word completely wrong in the first place. Combine that with a phone's personalized dictionary and you get completely authentic, utterly hilarious and thoroughly unrepeatable autocorrects. Our machines are smart, but they're not smarter than we are dumb … yet. Hooray!

SO WHAT'S IN IT FOR YOU?

It was tough to narrow down what to include in this edition, there are that many hilarious texts that come to DYAC and are book-worthy. In the end, I had to go with the best of the breast. Of the best. The criteria turned out to be simple: every single entry included had to have made me laugh hard enough to deem it re-readable, and those are the ones that made the cut.

Now without further ado, on with the auto carrots!

Family Time

Awkward exchanges
with your nearest
and dearest

Mom's New Phone

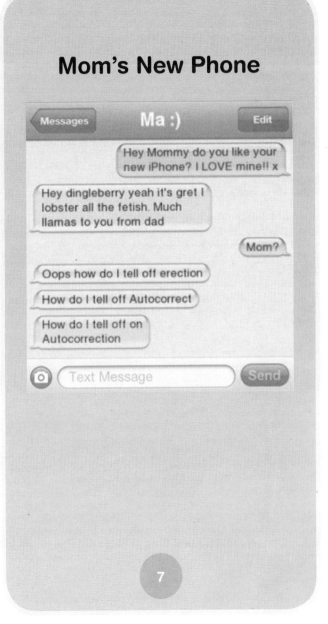

A Great Sale

Baggins

I was a little confused.

Jan 27, 2012 8:45 AM

I think our son is a hobbit. He is eating a third breakfast. Perhaps we should rename him dildo baggins

No.....bilbo bilbo bilbo! Not dildo!

I should have went with frodo!

That is awesome.

Not awesome. Horrifying.

Text Message Send

family time

9

family time

Blood is Thicker

Mom

Messages | Edit

have you talked to your sister yet?

No I'm still mad at her.

Sarah you need to let this go. she is your only sister

and blood is thicker than Worcestershire

I will remember that next time I'm eating steak!

oh damn this phone. I meant water. now call her.

Text Message | Send

Camo Pants

Jan 28, 2012 9:30 AM

Guess what I found for baby hayley!!

Crotchless panties

WHAAAATTTT?!?!? U want ur 6 mo. Old GRANDDAUGHTER to wear CROTCHLESS PANTIES?!?

Oh my gosh, I meant CAMO PANTS... ew, how did that end up as crotchless panties?

Ugh, I was about to disown u!! Lol

Text Message · Send

13

family time

family time

Mom's on Facebook

Messages · **Mom** · **Edit**

Lips are sealed.

Apr 27, 2013, 9:47 AM

Can I do laundry today

Apr 27, 2013, 10:58 AM

Yep

Apr 27, 2013, 8:09 PM

How do you delete a pussy on Facebook?

Do what mother

Lol. Sorry, i really need to start wearing my glasses when using this phone. I meant how do you delete a post on Facebook. Good grief......

Text Message · Send

18

Happy Mother's Day

Stacy

May 13, 2012 10:36 AM

> Happy mothers day shitstain! Luv luv xoxo

> OMG I meant sweetness! Way to ruin your day, huh? Lol I'm so sry

> I love you!

> I'm a horrible person and going to hell

Shitstain....Hahaha! Well happy mothers day to u too fuckface...LOL... Love u too!

> Lmao touché!

DAMN YOU, AUTOCORRECT!

20

family time

21

Johnson

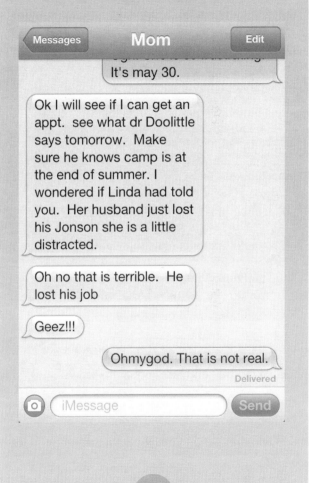

It's may 30.

Ok I will see if I can get an appt. see what dr Doolittle says tomorrow. Make sure he knows camp is at the end of summer. I wondered if Linda had told you. Her husband just lost his Jonson she is a little distracted.

Oh no that is terrible. He lost his job

Geez!!!

Ohmygod. That is not real.

Delivered

iMessage   Send

family time

23

Mary Poppins

something soon okay I miss you already okay bye

Mommas Cell   8:24PM
- The kitchen is super calla fragile list stick XP all the oh shit

Mommas Cell   8:25PM
- Ha ha Suprcalifragilisticexpialocious!! From Mary Poppins!!

Me   8:29PM
- ✓ LMFAOoOooOoooO hahahahahahahahahahaha

DAMN YOU, AUTOCORRECT!

24

What Now?

E. as u can see. No lol
can t figure it out foo much
so. As. Snartnea

No can t figure it out too
smart for me thinking
about taking it back and
we. Exchanging it for a
simple one son t know
how to erase a letter when
I make a
Istakee thank u for calling
back. Web going FTP to
call u all weekend got
agley today and every
mondaty Olds woman
son t news this phone lik
love u mom

> Is this in english? Lol

Jul 16, 2012 10:45 AM

My d. Fingers are too bu.
bio. Big for the keypad. T
his is pig Latin no book
came with it poi. Piece if
paper that I can't read with
a magnifying glass it is an
iPhone what was I
thinking. Lol 143

📷 (Text Message) Send

family time

25

Nap Time

Oh it's ok, I went to sleep on the blow up mistress, lol

Lolol...read what u just sent   lolol

Ugh blow up MATTRESS Lolol! Stupid auto correct!! Haha, sorry mom :))

Lol I knew what you meant. Didn't think u had a blow up friend. lol

Lol!! That's too funny! Hahaha!! Swear I meant mattress, no blow up mistresses here, lol!!

Text Message   Send

27

family time

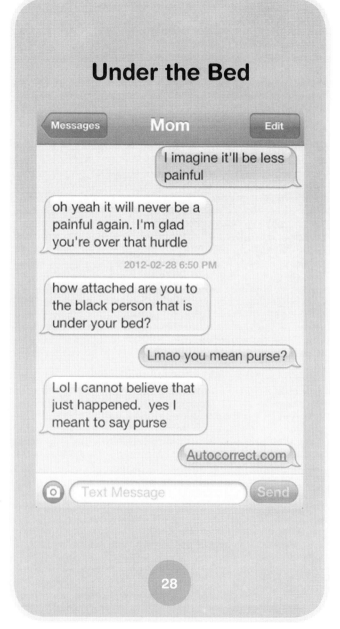

Under the Bed

Messages **Mom** Edit

I imagine it'll be less painful

oh yeah it will never be a painful again. I'm glad you're over that hurdle

2012-02-28 6:50 PM

how attached are you to the black person that is under your bed?

Lmao you mean purse?

Lol I cannot believe that just happened. yes I meant to say purse

Autocorrect.com

Text Message Send

DAMN YOU, AUTOCORRECT!

28

Photographer

Messages · Aunt Jenni · Edit

> I was thinking. I wanna be a prostitute photographer.

Nice

> PROFESSIONAL! Not prostitute! Damn you autocorrect

> You're telling everyone now aren't you

I already did

Text Message · Send

29

Snatch and Coffee

MOM <3

cardboard should be on the window ledge.

Are you coming back?

Yes with snatch and coffee

Yes! I love snatch. You're the best mom! And coffee? Good idea, I'll need the energy.

Jan 25, 2012 6:46 PM

OMG! I meant snatch!

Snacks*

Bring both.

Shut up!

DAMN YOU, AUTOCORRECT!

Uncle

Uncle dickhealer wants you poppin checkin grandpa when you get done with whore

None of the was right at all. Uncle denny wants you to stop and check on grandpa when you get done with work

Mar 13, 2012 5:30 PM

Huh

Text Message Send

DAMN YOU, AUTOCORRECT!

Unicorn Titties

family time

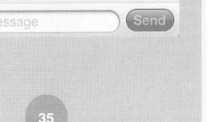

Crazy Mom

Waiting for prescription to be ready.

You are all you got it for me is another Bengali you all Landerset and many more on the wall and her mom and her mom I'll balance for my back and you are gay

My phone made this message for you. I'm not crazy...

Maybe you're not, but your phone is! What was that???

Text Message   Send

family time

35

Getting Testy

Don't forget your photo shoot with Jonas today

Oh yea. What time is it

3:30, address again is 777 West End Ave

Good luck and make sure he gets at least 10 different testicle shots

Testicle shots? I thought this was an engagement photo mom lol

Testicle shots!

Oh you know what I mean. Just get them! You only get married once. I HOPE.

Text Message   Send

family time

Halloween Surprise

Happy halloween dad! Are you dressing up :)

Nah, I'm way too old to be answering the door in a cockring

Ha! Costume! What a bad auto correct. Fat finger syndrome.

Yikes dad! Big difference!

Text Message   Send

DAMN YOU, AUTOCORRECT!

38

2

Getting Friendly

Autocorrect fails that will make or break your friendships

What's That Smell?

Holy fuck it smells so strong of colon upstairs 😑

Strong colon? Lol

-_- lol cologne

Yoda Ass-Kicking

My allergies are kicking my assss Yoda

Yoda lolol

Wtfff yoda? lol **today!

The allergies are strong with this one. Take claritin he must.

Lmfao I'm laughing so damn hard

getting friendly

Cats and Dong

42

Crotch Hole

Tawnya

Messages   Edit

I need your mom to sex my crotch. I have a hole.

Omg. Sew! Sew my crotch. In my pants. This whole text is awkward...

Feb 21, 2013, 11:48 AM

Lmao! My dad just read it and thought the same thing

You showed your dad?! Ha. Hi my name is tawnya and i'm awkward.

Feb 21, 2013, 1:01 PM

I am still laughing from this text

Text Message   Send

getting friendly

43

Don't be a Fool

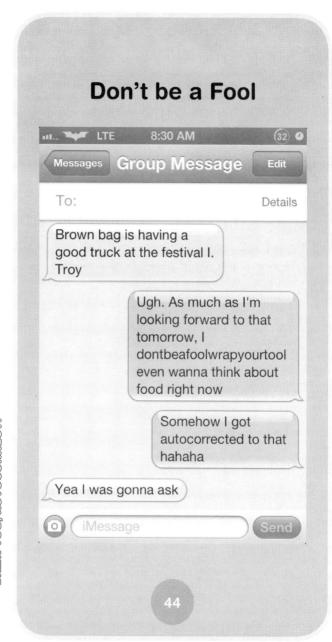

Group Message

Brown bag is having a good truck at the festival I. Troy

Ugh. As much as I'm looking forward to that tomorrow, I dontbeafoolwrapyourtool even wanna think about food right now

Somehow I got autocorrected to that hahaha

Yea I was gonna ask

44

Facial Hair

getting friendly

45

Freudian

46

Granny Balls

getting friendly

Harry Potter

2011-12-19 10:39 PM

Let me know if you dumbledore

Wtf? I have no idea how remember changed to dumbledore

Lol ya I was a little confused lol. Will do

Apparently my phone is a Harry potter fan

DAMN YOUAUTOCORRECT.COM   Send

DAMN YOU, AUTOCORRECT!

48

Llama

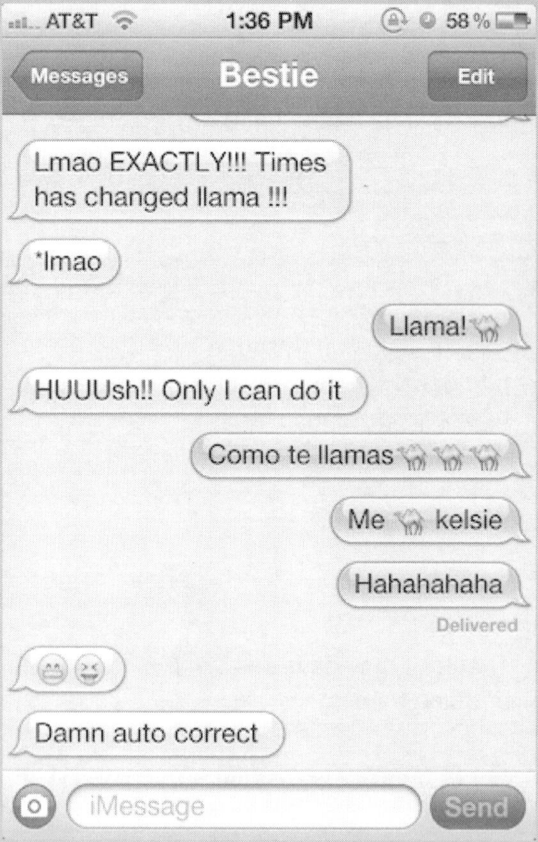

AT&T
1:36 PM
58 %
Messages
Bestie
Edit
Lmao EXACTLY!!! Times has changed llama !!!
*lmao
Llama!
HUUUsh!! Only I can do it
Como te llamas
Me kelsie
Hahahahaha
Delivered
Damn auto correct
iMessage
Send

Man Hair

Clear All   **Kristie**   Cancel

...ard of two. We will see.

You mean you're not going to hand stitch a blanket and baby clothes? Ha

I was going to make a blanket with my man hair but I thought that would be taking things to far.

Ew thats disgusting...

BARE HANDS!!! Not man hair

DAMN YOUAUTOCORRECT.COM ward

getting friendly

51

McCarthy

So, how goes it?

May 2, 2013, 2:16 PM

Can you call me when you get McCarthyist

That was supposed to say in the car I don't think you're a communist

May 2, 2013, 3:59 PM

Sure!

📷 | Text Message | Send

DAMN YOU, AUTOCORRECT!

52

Kewlcumber

getting friendly

53

getting friendly

Nuns

Only if there are no puppet people near.

Lol. It's so hot. Like there are a thousand burning nuns in the sky.

I don't know how you'd launch that many nuns into the sky. Especially if they are on fire.

What?? ?

I dunno. You're the one burning nuns.

Damn phone. SUNS. Not nuns. Lol!

DAMNYOUAUTOCORRECT.COM

Pride

getting friendly

Vagasil

Messages **Lauren** Edit

Ft x

5 May 2013 11:35

Fancy going up to teach with me to get som Vagasil

Vaseline

Auto correct ;)

Hahahhahahhahah

Delivered

Send

DAMN YOU, AUTOCORRECT!

58

Sexy Klingon

getting friendly

Show Me the Tortoise

.ıll AT&T 📶   7:37 PM   🔋 30% 🔌

Messages   **Debs**   **Edit**

| Call | FaceTime | Contact › |

Text Message
Jan 26, 2012 7:32 PM

> Girlllllllllll show me your tortoise!!!!

> *titties

> *TATTOO GOD DAMNIT

> And hello :) hahahahaha

Hahahaha, that's the best auto correct fail I have ever received.

And there's pictures in my winter break album on Facebook!

📷 **DAMN YOUAUTOCORRECT.COM**   Send

60

So Old, You're Prehistoric

ooo. AT&T 3G   11:28 AM   72%

Messages   **Group MMS**   Edit

To: Details

Text Message
Jun 13, 2012 11:25 AM

Hey ladies progressive dinner is tonight at 545 an it two dollars hope to see your pterodactyl faces there!

Wow I wrote pterodactyl not pterodactyls

B E A U T I F U L! I wrote beautiful none of you look anything like a dinosaur!

DAMNYOUAUTOCORRECT.COM   Send

getting friendly

61

Sparkly

Messages (1)   **Matt** ████████   Edit

for my moms birthday in a few months were gonna get her a sparkly orgasm bucket :)

OMG. how did that just happen?! pandora bracelet ******

I'm laughing my fucking ass off hahahahahahahaha

Read 12:43 AM

LMFAO. THATS GOING ON FACEBOOK AND INSTAGRAM. 😂

omg that's seriously the funniest thing ever.

DAMNYOUAUTOCORRECT.COM   Send

DAMN YOU, AUTOCORRECT!

62

Stick with Neutrals

getting friendly

DAMN YOU, AUTOCORRECT!

Stray Tanning

ıll. Verizon 📶   8:38 AM   42% 🔋

Messages   G   Edit

May 3, 2013, 8:20 AM

Ever done stray tanning?

Like, randomly tanning stray animals?

Spray tanning.
😂😂😂😂

😊 .... Yep, I have done several times

I'm seriously about to pee my pants.

I'm sitting in orthodontist waiting room trying to stifle a giggle

Might have to send that one in to damnyouautocorrect.com

I think it would make it

📷 iMessage   Send

Ovarian

5 minutes ago

I'm ready for thong welching to be ovulating.

What!?

*the weekend to be ovaries

*OVARIAN

*OVER! WHAT IS WITH MY PHONE AND THE FEMALE REPRODUCTIVE SYSTEM?!

getting friendly

Thunder

I love thunder so much

I know me too! Except when it makes my dog bark, so usually only when I'm outside

My dad barks too, we have to comfort him

I'm so sorry, I imagine his manly vocal chords cause much more commotion than a small dog

Jul 7, 2012 9:33 PM

OH DID YOU MEAN DOG

YEAH NO DUH

Text Message Send

DAMN YOU, AUTOCORRECT!

66

Tomato Warning

Messages   **Meagan**   Edit

tomato, where is it?

> I've never seen a tomato in Alabama.

The radio said theres a tomato warning tho

> Meagan. I'm pretty sure tomatoes aren't that deadly.

Nvm the tomato is in tuscaloosa, we already got outta there

> Better keep moving or the tomato might ketchup with you.

DAMN YOUAUTOCORRECT.COM   Send

getting friendly

67

Love in the Time of Text Messages

When couples text

Haircut

love in the time of text messages

How Do You Take Your Coffee?

Elfish Behaviour

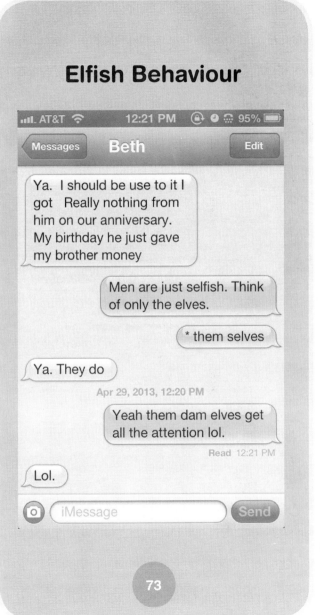

love in the time of text messages

Grandpa's Advice

Messages **Kristen** Edit

I think grandpa wants us to have the oppertunity live together now that we r engaged cuz grandpa always said try the hoes on before u buy them.

Omg I said hoes!! Ahaha I ment shoes laughing my ass off!!!!!

Haha

Lmao! Yes try those hoes on before you buy them!!! Bahahahaha ha

Apr 14, 2013, 7:31 PM

Subject

iMessage

Send

DAMN YOU, AUTOCORRECT!

74

love in the time of text messages

Livin' and Lovin'

Call   FaceTime   Contact ›

Text Message
Jan 27, 2012 9:28 PM

How's loving with your bf? Do you enjoy it or does it become a hassle sometimes? Lol

To b honest hes kinda big and it hurts a lot of the time so I don't like doing it

Omg I totally meant living haha stupid auto correct

 DAMNYOUAUTOCORRECT.COM   Send

77

love in the time of text messages

Sexy Linguini

Text Message
May 8, 2013, 11:53 AM

When u go all the way, guarantee you'll want sexy linguine

Lmfao that would be lingerie!

Fck me

Hahahha

Let u knw in few

Right?

One of my better autocorrects

Text Message · Send

DAMN YOU, AUTOCORRECT!

78

Squirter

not! Haha.

Yes you were!!! I'm so recording you

I wasn't! Flinched and moved a bit, but not squirming. I'm not a squirter!

Squirmer! Motherfucking phone! Hahaha

Hahahahaha omg perfect autocorrect :)

Apr 26, 2013, 5:45 PM

Yes well....I've got nothing. Haha

love in the time of text messages

Staple Relationship

iPod 🛜   7:41 PM

News Feed   **Sebastian**   ℹ️
active now

So im not that bad of a person? And i have to admit i did meet you under stalker like circumstances...

Yeah. I mean I've had a lot of staplers. At first you were no different.

Staplers?

Ha! Autocorrect. I meant staplers!

Stalkers!

ROFL. and awww you really do care. ☺

+   Send

Respect

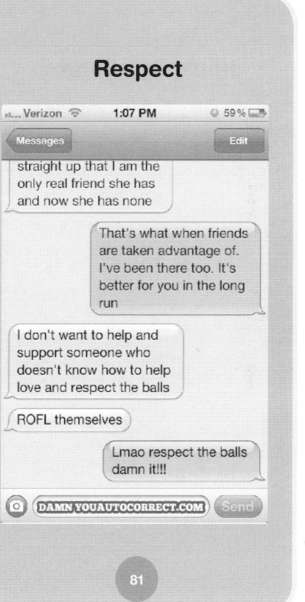

Messages   Edit

straight up that I am the only real friend she has and now she has none

That's what when friends are taken advantage of. I've been there too. It's better for you in the long run

I don't want to help and support someone who doesn't know how to help love and respect the balls

ROFL themselves

Lmao respect the balls damn it!!!

DAMNYOUAUTOCORRECT.COM   Send

love in the time of text messages

Hiding the Bodies

Messages **Rebecca new...** Edit

Can I wrap you in bubble wrap yet???

> Lol, once I get rid of this corpse then I can have my lymph node tested

> Boris mp

> Virus not corpse

I was hoping that was what you meant I don't have the money to fly over and help you hide a body :)

> Lol Siri will tell me where to hide it

📷 iMessage Send

love in the time of text messages

Love-Hate

Circus-Sized

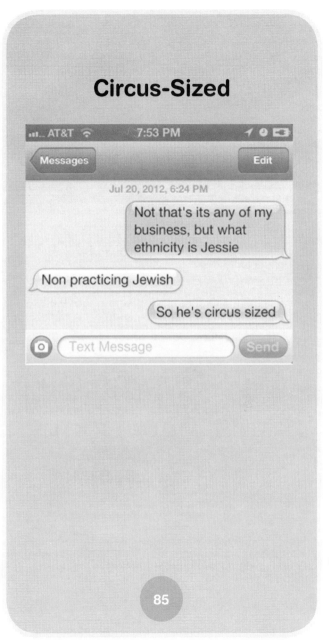

love in the time of text messages

Cow Strike

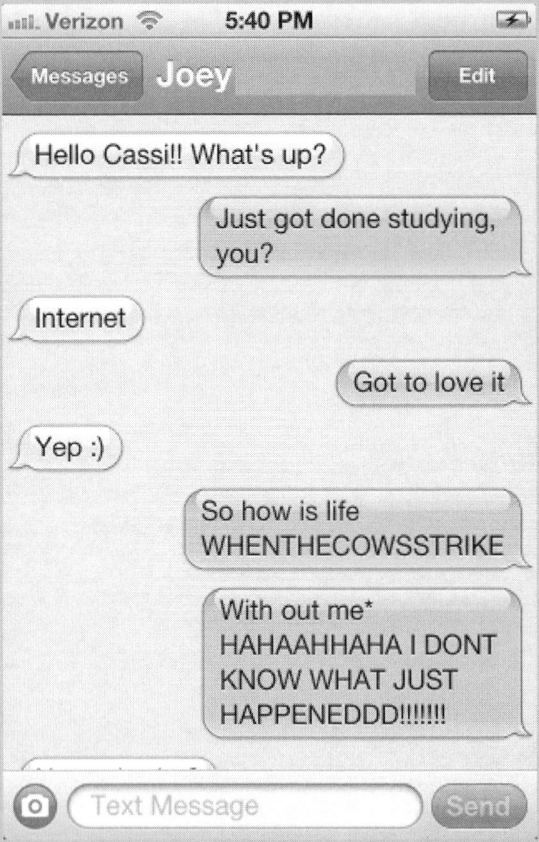

Verizon
5:40 PM
Messages
Joey
Edit
Hello Cassi!! What's up?
Just got done studying, you?
Internet
Got to love it
Yep :)
So how is life WHENTHECOWSSTRIKE
With out me* HAHAAHHAHA I DONT KNOW WHAT JUST HAPPENEDDD!!!!!!!
Text Message
Send

love in the time of text messages

My Hero

One more or 2 more?

2012-06-15 11:00 PM

One more after this

Ok :)

2012-06-15 11:22 PM

Love you

Your my best friend, my rock, my knight and shining armpit.

Omg armoir!! Hahahaha

Fail
Delivered

Epic fail

DAMNYOUAUTOCORRECT.COM

love in the time of text messages

DAMN YOU, AUTOCORRECT!

Nickelback

Messages   Adam   Edit

I love all types of music but 50s and jazz.

lol jazz is so bad, i hate it too

don't hate me but i've been really into necrophilia lately.

That is sick and not funny.

Please don't txt me again

holy shit wait

**nickelback that was the worst auto correct

DAMN YOUAUTOCORRECT.COM   Send

90

Oblong

love in the time of text messages

Under Attack

Sabrina

2012-05-28 9:06 AM

Morning babycakes! xo

2012-05-28 9:48 AM

Worst night ever

Aww why??

Woke up with penis attack and diarreah.

Panic

I think i ate a bad dick

Ugh dinner

I like the idea of a penis attack better!

love in the time of text messages

So Much Sausage

love in the time of text messages

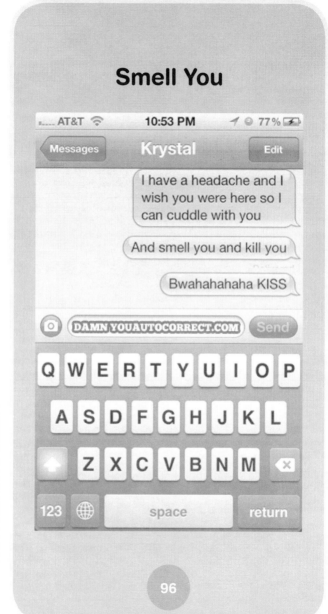

love in the time of text messages

Love Is

you

Apr 6, 2012 11:11 AM

I love you too babe

I hope you're feeling better babe. I haven't slept very well the last 2 nights either. I miss sleeping next to you and just being able to touch you or kiss you or lay my wad on your shoulder.

Head***

Wow.....it was so sentimental until auto correct fucked it up!!

DAMNYOUAUTOCORRECT.COM

Kitchen Nightmare

Food and drink texts gone wrong

50 Shades

Messages **Steph** Edit

> Stopping at holiday to get something to drink..want anything?

Dirty choke please

> Yea not so sure they sell that at the gas station but I can certainly check ;p

Omg diet coke damn auto correct lol but maybe deep down I wanted a dirty choke haha jk

> Have you been reading 50 Shades again....

Nope S&M R US catalog

Text Message Send

kitchen nightmare

101

A Chill Night

getting black out drunk to celebrate the first day over 87 degrees

Yea...not up for blackout...I might however be up for dinner and a couple drinks or something. I've got steaks masturbating if u guys are up for a chill night

HahHHHah marinating! My steaks aren't masturbating

May 10, 2013, 2:47 PM

Hahaha, hello damnyouautocorrect.com!

📷 (Text Message)   Send

DAMN YOU, AUTOCORRECT!

102

Baked Photato

kitchen nightmare

Bitch Goblet

AT&T 3G 5:40 PM 73%

Messages <3 Babe Edit

Mashed potatoes or bitchgobblet potatoes tonight?

What the fuck????

Omg...butthurt potatoes

Bitthurt

Delivered

Ughhhhh!

Buttered

Text Message Send

104

kitchen nightmare

Buttersex

Messages   **Kyle**   Edit

Today we were talking about childhood favorite ice cream flavors at school and somebody said buttsex penis! I thought of you

OMG DON'T READ THAT

Butter pecan*

How very in springhare

Inappropriate

Kill me

I'm not really sure what to say

DAMN YOUAUTOCORRECT.COM Send

Caffeine

Sammi
1 hour ago

Why on earth would I get another energy
drink when I've already had a butt load of
caffeine today? Oh yea cuz im a caffeine a
holocaust that's right lmao

Like · Comment

👍 Tator likes this.

Tator
>:O
1 hour ago · Like

Sammi
Oops caffeine a holic not holocaust
lol damn you phone
1 hour ago · Like · 👍 1

Write a comment...   Send

107

kitchen nightmare

Cameltoe Tea

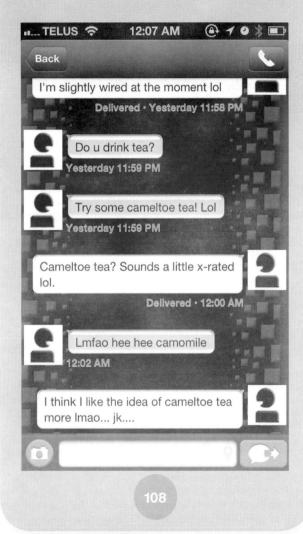

DAMN YOU, AUTOCORRECT!

kitchen nightmare

109

Caesarean

Messages   **Chris**   Edit

2012-08-19 1:37 PM

Hey. What would u like for dinner? Want me to pick stuff up?

2012-08-19 2:55 PM

no Chinese please. if you want any else thats ok

Oh no I wasn't thinking takeout. Want chicken cesarean wrap? I can cook it up when I get home?

2012-08-19 3:22 PM

oh ok

2012-08-19 3:41 PM

Send

DAMN YOU, AUTOCORRECT!

110

A Master

Messages   **Outgoing**   Edit

To: Details

24 May 2012 17:38

It's soo hot I can't even get a breath x

25 May 2012 17:16

Had a good afternoon at work we had a master cock licker in showing us how to make chocolate it was a bit meaty though in the heat, is it roasting up there today? X

Omg I meant chocolatier not cock licker and Melty not meaty omg lol

DAMNYOUAUTOCORRECT.COM   Send

kitchen nightmare

111

Way To Start The Day

Messages **Katie** Edit

> Yeah my morning was shitty...and the rest of the day was hot...Monterey is good.

> This morning I choked a goat and pissed in my coffee.

Lol!!

> LOL NO...I choked on TOAST and SPILLED my coffee...fuckin spell check...

I like the other one better!!

> You would...hippy...

DAMN YOUAUTOCORRECT.COM Send

DAMN YOU, AUTOCORRECT!

112

Cosy Sausage

come. We can make a mix match

I have chicken, sauce, that sausage thing eye

That what? I have veg and bacon

Delivered

I have a cosy sausage

Cold

kitchen nightmare

Double Dipping

Good idea. I never thought of doing individual cups of dip. Eliminates double dipping!

Omg those look good!!! And that dip idea is awesome! Hermaphrodites like my dad would actually eat the dip! Lol

Omg not hermaphrodites lol I typed germaphobes

Delivered

OMG!!! That was funny!!!

iMessage   Send

kitchen nightmare

kitchen nightmare

117

Freshly Ground

DAMN YOU, AUTOCORRECT!

118

Got Beef?

> Babe I don't feel like cooking tonight. Can you bring home human beef?

WTF Beth? I'm in a meeting. Human beef? Are you high?

> Hunan beef! The place that just opened on 7th ave!!!

> I'm laughing so hard I almost puked.

Jesus! I just laughed out loud and could possibly get fired now. :) Order your human beef. I'll pick it up at 6. Love you

DAMNYOUAUTOCORRECT.COM   Send

119

kitchen nightmare

Leftovers

kitchen nightmare

Bell Lepers

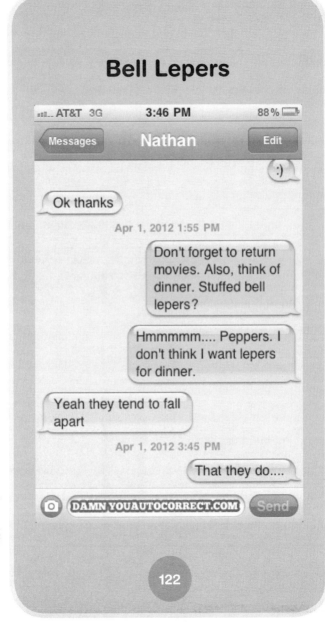

122

kitchen nightmare

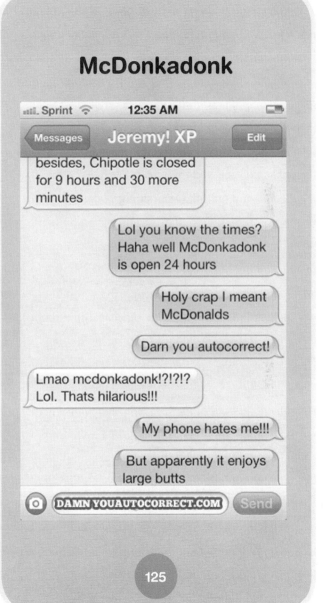

McDonkadonk

besides, Chipotle is closed for 9 hours and 30 more minutes

Lol you know the times? Haha well McDonkadonk is open 24 hours

Holy crap I meant McDonalds

Darn you autocorrect!

Lmao mcdonkadonk!?!?!? Lol. Thats hilarious!!!

My phone hates me!!!

But apparently it enjoys large butts

📷 **DAMNYOUAUTOCORRECT.COM** Send

125

kitchen nightmare

Got Milk

Koodo 3G   11:54 AM   69%

Messages   **Brandon**   Edit

Can I ask a favor of u

What would that be, going back to work in about 20 minutes

After work can u bring home a 4 litter of white whore milk we are all out and I'm craving it badly

White whore milk? Not sure where to get that but I'll try 😜

Hahaha duck I ment white homo milk

Fuck not duck shit

DAMNYOUAUTOCORRECT.COM

126

DAMN YOU, AUTOCORRECT!

Bad Priests

I made a crisp tonight out of some priests that were about to go bad. It was delicious!
Yesterday, 5:35 PM

Were they Catholic?
Yesterday, 5:40 PM

Huh?
Yesterday, 5:41 PM

Pears.
Yesterday, 5:41 PM

Omg.
Yesterday, 5:41 PM

kitchen nightmare

Secret Ingredient

whaaaat? (:

03/15 1:49 PM

Im making chocolate chip cookies with period in them(3 ((:

03/15 1:53 PM

period??? hahaha

Bahahahahhahahahahahah ahhahahahahahahhahahahah ahahahahhaahahahahhahahaha hahahahah OREOS!!! I meant oreos$!!

bahahahahahahahahaha i just peed!!!!

iMessage   Send

DAMN YOU, AUTOCORRECT!

130

kitchen nightmare

STFU

Mommy

Dec 24, 2012, 3:52 PM

Hey sweetie, I got stuff for Christmas dinner

Guess what I got for you

STFU MUSHROOMS!!!!

Omg what?

Stuffed Mushrooms***

Omg

Mom this is going on damnyouautocorrect.com

Delivered

132

Subway

vodafone UK 3G 13:12   97%

Like

Mm nothing better than the smell of a free 6" sub wafting out of my vag!!

1 hour ago near Swindon

👍 2 people like this.

I hope u ment to say bag??!!

1 hour ago · Like · 👍 1

ROFL!! Obviously I meant my vag, cause that's where I store all my subs lol I meant bag yes x

1 hour ago · Like · 👍 3

Hahahahaha!!!!! Too funny! Xxx

1 hour ago · Like · 👍 1

DAMNYOUAUTOCORRECT.COM   Send

133

kitchen nightmare

Hershey Kiss

Peanut Bastard

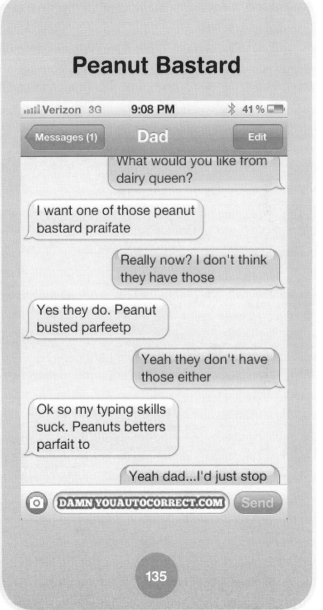

kitchen nightmare

DAMN YOU, AUTOCORRECT!

5

That's Entertainment

TV text fails

Batman

that's entertainment

139

that's entertainment

Game of Thrones

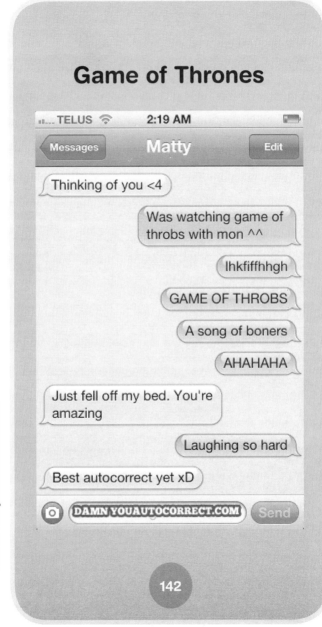

Cloudy Balls

that's entertainment

P.O.T.T.E.R

Messages **My Boi Scott** 💀 Edit

Yeah lol. You're silly. And dumb. And that's good. And I know silly boy! Lol

Mar 18, 2012 15:51

Lol. You're silly too! And I didn't know about the hairy penis weekend!!! But I have to go to church so I can't watch them all :(

*HARRY POTTY!

*P.O.T.T.E.R.

You did not just say harry penis.

I didn't mean to!!!!

📷 DAMN YOUAUTOCORRECT.COM Send

Irwinman

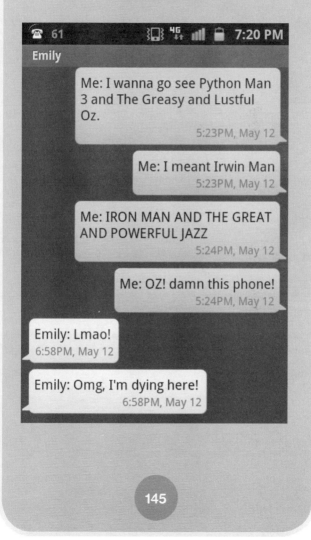

Emily

> **Me:** I wanna go see Python Man 3 and The Greasy and Lustful Oz.
> *5:23PM, May 12*

> **Me:** I meant Irwin Man
> *5:23PM, May 12*

> **Me:** IRON MAN AND THE GREAT AND POWERFUL JAZZ
> *5:24PM, May 12*

> **Me:** OZ! damn this phone!
> *5:24PM, May 12*

> **Emily:** Lmao!
> *6:58PM, May 12*

> **Emily:** Omg, I'm dying here!
> *6:58PM, May 12*

that's entertainment

145

Love a Baboon

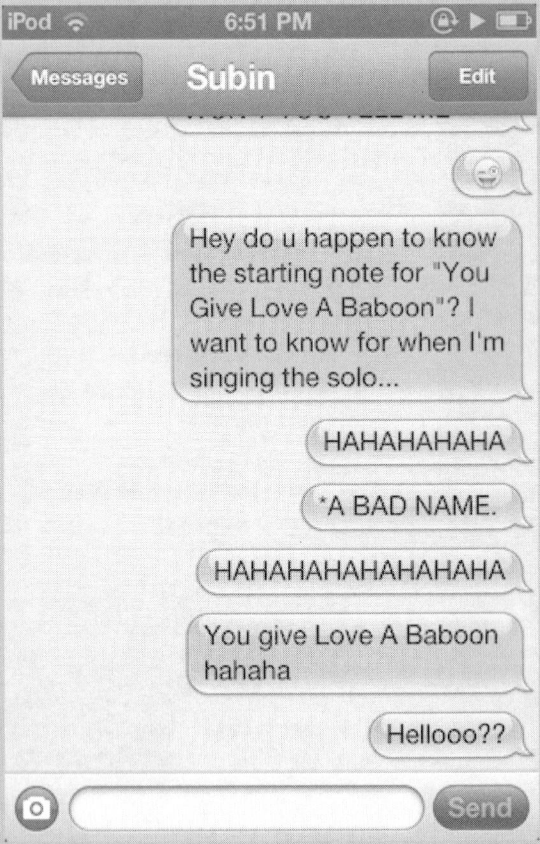

that's entertainment

Moves Like ...

148

Movie Night

> Nice. We are watching les mozzarella
> ✓ Delivered

> Les miz
> ✓ Delivered

What the heck is that?

A French pizza movie

> Auto correct
> ✓ Delivered

that's entertainment

Stumperman

that's entertainment

10 Things I Hate About Carrots

Messages   Edit

Jan 23, 2012 6:09 PM

I am staying up to watch a movie. Contact, Pirates of the Carrotpenis, or 10 Things I Hate About You....trying to figure out what im in the mood for. Onl

y one more kid to public down, then I'm curling up with a blanket. Fairly easy night!! Have a safe trip!@

Carrotpenis?!?

Huh?

Holy shit nuggets! Lmfao!

📷 DAMNYOUAUTOCORRECT.COM   Send

153

that's entertainment

Wheel of Fortune

that's entertainment

Bach

Bill Clinton

store.

Sep 5, 2012 5:33 PM

RAPE BILL CLINTON

Wow. That was supposed to be tape.

Hahahahahaha

Tape Bill Clinton. Do not rape him.

I'm almost certain you can get in trouble for that.

I don't think he would put up that big a fight.

Good point.

DAMNYOUAUTOCORRECT.COM   Send

157

that's entertainment

Personal Jesus

Jessi

Text Message
Mar 3, 2013, 4:06 PM

This is Jesus this is my new number :-)

Jessi

Mar 3, 2013, 5:57 PM

I was shocked that Jesus felt like I was a close enough friend to give me his number...

DAMNYOUAUTOCORRECT.COM

Star Wars

Eh. Same old same old.
Need beer money :-P
How was the checkup?

May 14, 2013, 8:19 PM

It went great! Turns out I have retrograde Star Wars. So that is why my right ovary was scarred and twice as big. That is what was causing the pain.

How the hell did endometriosis turn into Star Wars? That is just unreal.

Apparently my uterus is the Death Star.

LMFAO

DAMN YOUAUTOCORRECT.COM   Send

DAMN YOU, AUTOCORRECT!

160

6

Taking Care of Business

In and out of
the office ...

Report

Nic   : You still at work

22:01, Feb 26

Me: Yes. At the vaginal stage of a very long lobster report. I'll beat you at the office in a minute .

22:03, Feb 26

Nic   : Look I know that you have been lonely lately but I'm not into stuff like that.

22:05, Feb 26

Me: Freaking Amoco correct ! I meant final stage of a long drawn out report!

taking care of business

163

Boobs

Busy Day

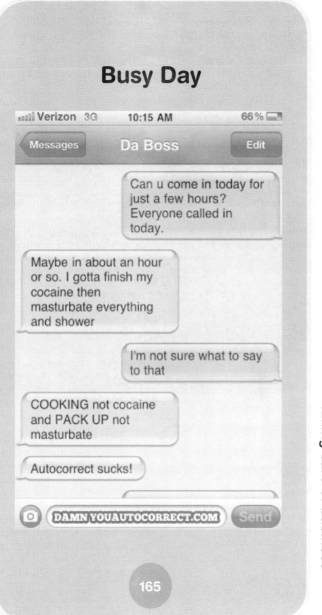

taking care of business

Pickup

Maggie L

Which door

I csn come to street

Coming

Oct 27, 2012 8:58 AM

I'm going to swing by homo depot before I get to the church is there anything I should pick up

Homo depot. That sounds like a different kind of store, lol!! Get me some chaps and a Barbra streisand album.

Read 9:05 AM

Co-Chair

Hi! is your dad still on the condo board of directors? I have a question.

Yep! He knows everyone here.

He's COCKGOBBLING.

Okay I have no idea what happened.

What!

I meant, he's cochairman. I bought a used Phone. This has been happening to me all week.

DAMNYOUAUTOCORRECT.COM Send

taking care of business

167

Dead Boss

> Dead Boss, The solicitor said he appreciated your offer, but he wasn't sure about car agora D. He said he would like to have a meeting with you a nod you should meet him in the entrance ball. Kind farts, Your secretory.

Read 14:18

Lindsay, Just to let you know that I am not dead, I do not know what you mean by car agora D, we do not have an entrance ball, and farts are not kind. You are in big trouble!

📷 Text Message | Send

DAMN YOU, AUTOCORRECT!

168

Studying Up

Hi Jen, this Craig Lee. I have results of Accounting midterm. You scored 74/100. Any questions call me. Thanks

Thanks Craig. I guess I will have to fuck ye even harder before the final. See you Tuesday.

Omg that was supposed to say "study" I'm sorry

It ok. I once asked my wife where the LSD was (meant kids). Have good night.

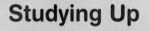

DAMNYOUAUTOCORRECT.COM   Send

taking care of business

Hedgehog

Cancel **Sick leave** Send

Cc/Bcc, From:

Subject: Sick leave

Dear Mr. ▇▇▇▇▇
I don't think I can come to work today, I woke up with a terrible hedgehog. As soon as I feel better, I'll go to work.

Q	W	E	R	T	Y	U	I	O	P

A	S	D	F	G	H	J	K	L

Heterosexual

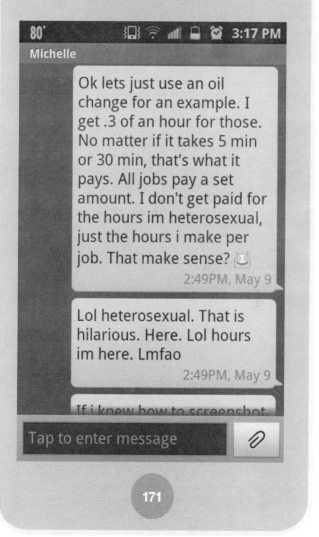

Michelle

Ok lets just use an oil change for an example. I get .3 of an hour for those. No matter if it takes 5 min or 30 min, that's what it pays. All jobs pay a set amount. I don't get paid for the hours im heterosexual, just the hours i make per job. That make sense?

2:49PM, May 9

Lol heterosexual. That is hilarious. Here. Lol hours im here. Lmfao

2:49PM, May 9

If i knew how to screenshot,

Tap to enter message

taking care of business

Jedi

Messages   **Amanda**   Edit

I swear to Jedi if I have to go into work early I'm gonna be pissed.

Wtf? Jedi? I meant god. Apparently my phone likes star wars?

holy jedi father of luke!!!

DAMN YOUAUTOCORRECT.COM   Send

DAMN YOU, AUTOCORRECT!

taking care of business

173

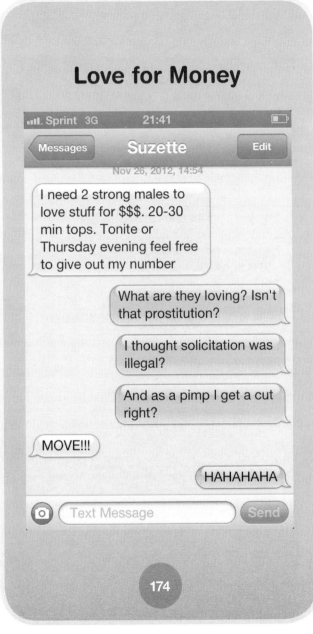

DAMN YOU, AUTOCORRECT!

Reproductive

Messages **Kyle** Edit

So I woke up and I thought I was going to be reproductive but then I fell on my bed and just wasn't...

Crap

Really productive

Hagahagaahagagagagaga haahahahahaha

That's not funny

Youre right its hilarious

Screw you. That's not my fault

Of course it isnt

iMessage Send

175

taking care of business

Rocket Tickler

Oh I was going to ask do you know anyone who has a rocket tickler that I can borrow for a hole

Holy shit! NOOOO

That's not what I meant at all!!!!

Let me try again

Lmao!!! I cant hardly text I am laughing so hard.

Sep 25, 2012, 10:15 PM

A roto tiller that I can borrow for a while! Wow really iPhone! Oh my god

176

Sold

taking care of business

Bad Day

working on, and I've essentially been demoted and Lea Anne told me that in front of Janna, Kim, and tiana.

I'm sorry you had a bad day. You work so hard and you don't deserve that. Cumstains being what they are, as long as it doesn't come with a pay cut maybe it's a blessing to be back to selling.

Circumstances.

BAAAAAHAAAAAHAAAAA A!

DAMN YOU, AUTOCORRECT!

📷 DAMN YOUAUTOCORRECT.COM Send

178

Touching

The Usual Stuff

Messages   **Mike**   Edit

Call   FaceTime   Contact >

iMessage
Mar 5, 2013, 7:15 AM

Printer not working, I tried the usual stuff - unplugging, beating it, put my balls on it and prayed?

Hmm that usually works.

📷 iMessage   Send

DAMN YOU, AUTOCORRECT!

180

Carwash

Hey, do you think you could wash my cunt this weekend? I'm too lazy, but it's soooo damn dirty. Plus, you're way better at detailing than I am.

What the actual fuck???? Car, I meant carrrrr.

Omg, two fails in one text. Wrong Mike. I hate this phone already. So embarrassing.

Hahahahahahahaha

Bahahahahahahaha

DAMNYOUAUTOCORRECT.COM   Send

taking care of business

Whale of a Time

Messages   Edit

> So the kids school book fair is this week and i was reading the kids wish lists. Haven seriously wants a book called "whale foreskin in statue"

> I guess it's educational. So I shouldn't have an issue

> Holy mother of all autocorrects!!! I meant Why Florida became a State!!!!!

> Duuuuuuude!!!

That sounds nasty

DAMN.YOUAUTOCORRECT.COM   Send

DAMN YOU, AUTOCORRECT!

182

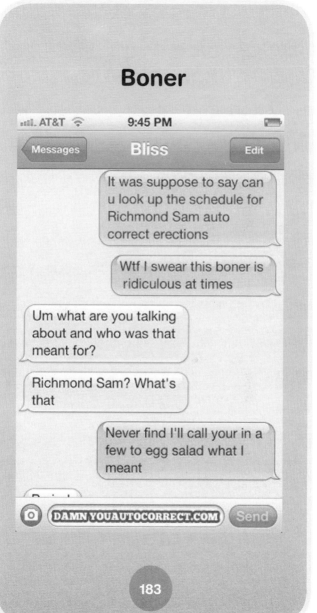

Boner

taking care of business

Drunk Texting

Messages   **Darell**   Edit

2013-05-13 8:04 AM

Wakey wakey

2013-05-13 8:19 AM

Shizuoka I'm amess. What is the possibility of coming in Kate today? I have no work today.

Wow autocorrect is being a bit of a bitch.

Reread your msg and laugh your ass off you drunk bastard lol. I'll text you with your new time for today

Your a beauty.

📷   Text Message   Send

DAMN YOU, AUTOCORRECT!

186

Ducking Around

taking care of business

Bear Porn

Massimo

Clear All · Cancel

Haha Ethan. Do u play ping pong a lot? Lol

No in Pe I just really suck and hit people in the face a lot

I'm really good at bear porn though

Lmaoooo bear porn

Hahahahahah

Beer pong fuck

Delete · Forward

7

TMI

Too. Much.
Information.

TMI

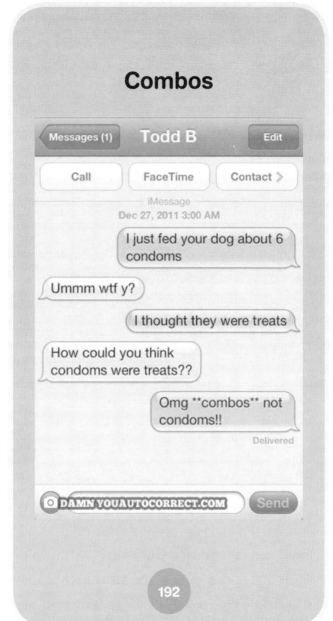

Corporal Punishment

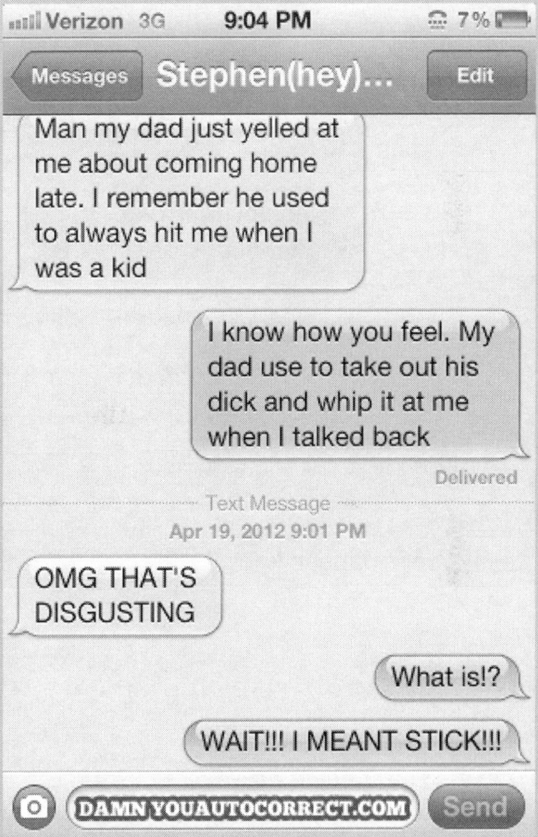

Crabs

Wanna come to m?

> Nahhh I have really bad crabs

> CRAMPS!!!!

> Oh ma gosh
> Delivered

Oh honey

What did I tell you about that...does Janson know?

📷   iMessage   Send

DAMN YOU, AUTOCORRECT!

194

Huge Hair

Ommmggg 😔 last night we had huge hair and it scared the crap outa me I totally shot up and freaked out by the hair banging on my window.

An yea it rained all day yesterday an last night but it hasn't really rained today

Lmao. Huge hair???? Dying laughing.

Going on dyac.

Lol! No huge hail*

Hair banging on your window? SNORT.

Hahahahhahahaha

TMI

Dressed

TELUS 9:23 AM 76%

Alison

I really like that...no matter what happens we will deal with it. That's so true. It's how I look at life now. I don't get dressed nearly as often.

Stressed!!!!

Don't get STRESSED nearly as often! Autocorrect hates me some days

That made my day!

Lmao!

I love having you in my life!

DAMNYOUAUTOCORRECT.COM

DAMN YOU, AUTOCORRECT!

196

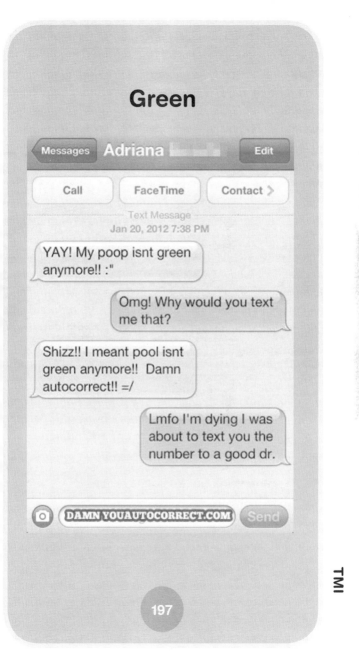

Green

Call FaceTime Contact >

Text Message
Jan 20, 2012 7:38 PM

YAY! My poop isnt green anymore!! :"

Omg! Why would you text me that?

Shizz!! I meant pool isnt green anymore!! Damn autocorrect!! =/

Lmfo I'm dying I was about to text you the number to a good dr.

DAMNYOUAUTOCORRECT.COM Send

TMI

197

Making Payments

Baybeee

I wish I had extra money...
I really wanna buy a lap
dance. Like a really
expensive one and make
payments.

Lol i dont even know

Oh what the hell. Laptop.
Not lap dance.

Ahahahhaha

Can you make payments
on a lap dance? Hahahaha

Apr 3, 2013, 9:17 PM

Hahaha idk but Im gonna
look into it

Fumigating

What's going on over there. Tom said you have ants?

Yes! It's horrible. They're in the kitchen and the dishwasher.

Mom's been running around fingering everything in sight and it's making the house smell so bad.

Now that's a bad mental image. I'm scarred for life.

Fumigating, not fingering! Bad, autocorrect! BAD!

Text Message   Send

TMI

Mouldy

Christine Hin...

1 May 2013 12:31

I've just found mould on my bum

1 May 2013 13:10

Sorry. I'm crying with laughter.

Have a word with the Doctor on Thursday.

???

Ah that should have said bun damn iPhone keys

Real tears.

Read 13:13

Glad I made you laugh x

200

New Pants

New Pants

ıɪ. Verizon 3G   10:03 PM

Messages   Edit

Call   FaceTime   Contact ›

Text Message
Mar 3, 2012 9:54 PM

My butt hole ripped and it looks happy

Wtf?!!! My pants!!! THe knee

LMAO OMG I CAN'T BREATHE

I was like "wtf?!"

Hahaha! I cant believe that

DAMNYOUAUTOCORRECT.COM   Send

TMI

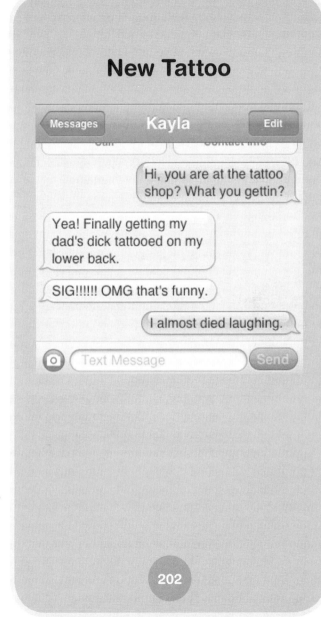

New Tattoo

Kayla

Hi, you are at the tattoo shop? What you gettin?

Yea! Finally getting my dad's dick tattooed on my lower back.

SIG!!!!!! OMG that's funny.

I almost died laughing.

Text Message

DAMN YOU, AUTOCORRECT!

202

Prego

Just saw. Is your sis prego?

6:31PM

Yeah bro I'm so stoked. I'm waiting on my mom to get home and we are going to take a shit over face time to celebrate

6:32PM

Taking shits over face time? Lol. Must be a family thing. Lol. That is awesome bro. Congrats. Is your sister in reno?

6:34PM

Wow fuck you auto correct. Shots lol

6:35PM

TMI

Pillow Talk

.ıl Sprint 📶 | 10:58 AM 72% 🔋

Messages **Caleb Mobile** Edit

My new bed is so fuck sharia's. I'm going to bed so early just so I can lay here and experience the comfort, pleasure

My new bed is so luck Sherry us...

Fucking Luck Sherry?

My new bed is so looks serious...

My new bed is so luxurious

📷 iMessage Send

204

Sore Legs

TMI

Superman

3G 11:55 PM 58%

Dammit. Proof it* force of habit

Are you later because I'm her and I'll come back and get you or are you why he hear how your mommy figured you coming or are I like to take your mom eat Indalinc like to what's going to sores or Superman Popsicle we'll get to go to you

The fuck...?

WTH where did all that come from!?!?

DAMN YOUAUTOCORRECT.COM Send

206

Suzy Scrotmaker

It haha

2011-12-22 10:53 AM

They don't call me Suzy scrotumlicker for nothing

Errr I mean Suzy homemaker

I'm still laughing so hard!!

Delivered

Haha A girls gotta do what a girls gotta do :)

DAMNYOUAUTOCORRECT.COM Send

TMI

Precious Time

T-Bag

TMI

Creepy

> it's on Sepulveda & mariposa near el segundo blvd. maybe meet there at 7?

8 is better maybe I'll let iknowitwasyou

u know* WTF DID MY PHONE JUST DO LOL

> hahahahahahahahaha

> oh my god your phone is creepy as fuck!!!

Delivered

DAMNYOUAUTOCORRECT.COM Send

Vapour Rub

TMI

April Showers

> I haaaaate April. This month is rainy and disgusting.

It will be over soon! April showers bring May flatulence!

> Oh because thats something to look forward to!

> Every month is flatulence month where I live. Have you met my husband?

lol

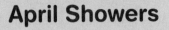

DAMN YOU, AUTOCORRECT!

Bite Balls Challenge

What do you do when you're nervous

Crack my knuckles and used to bite my balls but not any more

Oops meant my nails haha

Hahahahahahahahaha

Ok that is beyond hilarious

Wow imagine I could that

I will NEVER let you live that one down

I figured as much

DAMNYOUAUTOCORRECT.COM

TMI

Scented Candle

Hey girl does Costco sell scented candles?

No they don't

Well damn it. I'm looking for a candle made by Paula Deen it's scent is Blueberry Twat. . If you find it buy it for me and I'll pay you back.

I'm sorry blueberry what?

Oh well fuck me blind. Blueberry TART.

I was gonna say! Wtf mom blueberry twat?!?! Who wants that?!

Well not me! I don't care if its blueberry scented!

📷   Text Message   Send

DAMN YOU, AUTOCORRECT!

TMI

Piece of Shirt Phone

Messages Tyler Edit

Jul 9, 2012 3:15 PM

What's up?

Not much. I've just been sucking penis all day

That's good for you...

Delivered

Oh dear god shucking penis

PENIS

Peanuts!!!

Piece of shirt phone

DAMNYOUAUTOCORRECT.COM Send

216

Black Hole

No more black hole. You brought twats to someone's eyes!

Jesus Christ, phone. What kind of auto correct is that? I swear I did not do that on purpose. TEARS. TEARS.

Lol! Amazing!

That just went on FB, you're famous.

lol. And FACEBOOK SHALL KNOW FOREVERMORE

📷 iMessage   Send

217

TMI

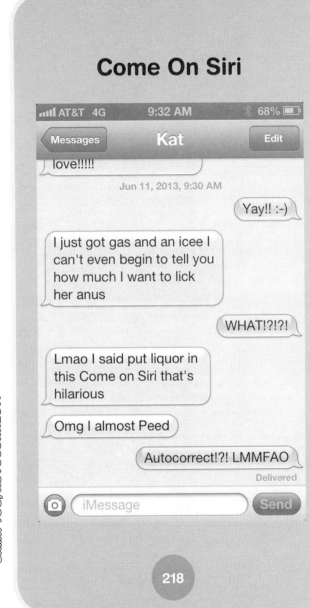

Who, What, Where?

Text fails anywhere, anytime

who, what, where?

Baptised

Baptised

Fido 3G
11:23 AM
95 %
Messages
Bryan N
Edit
Cousin night?
Karl Denny is getting vaporized, so I'll be at church.
Baptized
DAMN YOUAUTOCORRECT.COM
Send
Q W E R T Y U I O P
A S D F G H J K L
Z X C V B N M
123
space
return

Dead Walruses

Messages — Eric — **Edit**

My neighbor is so ghetto. His grass goes up to my calf. Mow your lawn mofo

yea the guy next to me wont take his xmas decorations down

its April and he's still got 3 dead walruses nailed to his house

Ew!!! Please tell me that was an autocorrect!!!

lol yea wreaths i meant ...

Wow I thought my neighbor was bad! :/

DAMNYOUAUTOCORRECT.COM

Send

225

who, what, where?

Extraterrestrial

Feb 22, 2012 2:44 PM

Hey I'm throwing up will you get extraterrestrials for me if she hands any out

Lol. Extraterrestrials? Yes. Feel better.

Lol

Extra poopturds

Awesome.

Damnit p a p e r s

Lol. I will get all of those things.

Delivered

DAMN YOUAUTOCORRECT.COM

Genitals

> 😱 More problems between those two....

> Maaaakay. I'm just getting in the Showa.

> Alright, I'm about to that point lol

> Ha alright!

> I'm wearing my genitals tonight! 😎

> 😳 OMG! **extensions!

Delivered

who, what, where?

Meth Homework

228

Inflate

who, what, where?

who, what, where?

Pompous

Bethanie

Text Message
Jan 27, 2012 8:19 PM

When you come to my house look for the pompous ass at the top of the driveway.

Pompous ass? Who might that be.

No, he's on the couch......I meant pompous grass.

Smartphone

Bahaha very smart phone ;)

DAMNYOUAUTOCORRECT.COM

who, what, where?

Alone Time

Messages   Edit

Can you please tell Robyn that I've got the gastro that everyone is getting at the moment, and that if I'm feeling better this afternoon, I'll try to make it in then. At the moment, I can't make it past an hour without cumming.

WTF. Vomming, I meant vomming.

11/09/2012 11:27 AM

Don't worry buddy. Robyn understands the importance of some 'alone time' every now and then.

DAMN YOUAUTOCORRECT.COM   Send

236

Smell Your Fart

You?

Feeling super tired and taking a power nap before grabbing food :P

I'm excited to see you smell your fart tomorrow

Holy fuck, epic google voice fail. *to see you SELL your ART. Wow. I'm mortified.

HAHAHAHAHA

LOLOLOL

07/26 6:15 PM

I. Am. So. Mortified. <_>

DAMNYOUAUTOCORRECT.COM

Send

who, what, where?

237

who, what, where?

Life in the Fast Lane

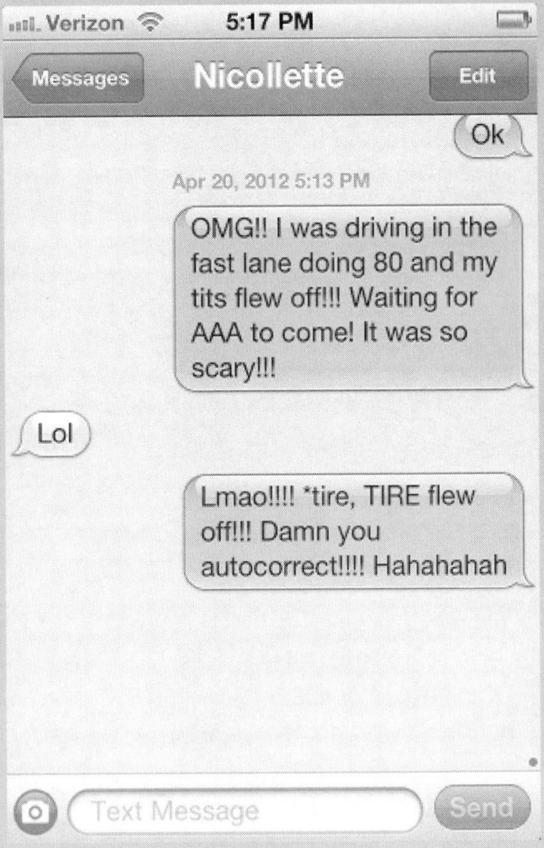

Art Form

who, what, where?

Brushing Up

You want to head over to standard?

I'm gunna leave my place really soon

Ok like what's really soon? Just gotta brush my uterus and stuff

Woooow teeth**

Ooh my god

Oh dear...

Best auto-correct ever?

DAMN YOU, AUTOCORRECT!

Happy Birthday Nan

Messages **Group Message** Edit

To: Details

iMessage
Apr 29, 2013, 3:39 PM

Mom

Would you be available this evening if Nan would like to ho out for dinner for her birthday?

Sure

But tell nan she doesn't need to ho out. I'm sure we can all chip in and give her some money

iMessage Send

who, what, where?

243

Duck

> The Easter bunny is frightening, but I think they all are. However, there is a cute setup where your kid can hold a baby dick, a lamb, or a real bunny!

Omg!!! Baby FUCK!

NOOOO!!!!!

Baby DUCK! duck!

WHAAAAAAAAAT

Geez woman!!!!

My autocorrect is evil!!!

📷 DAMN YOUAUTOCORRECT.COM Send

Golf Balls

who, what, where?

Happily Ever After

Messages   **Joe**   Edit

Oos?

Apr 28, 2013, 12:45 PM

We are in love at Arby's.

Line. This phone is so gay.

I'm happy for both of you

Couldn't you find a more private place for your man love?

Must have been some short love making

Apr 28, 2013, 1:02 PM

Very short. Yo know what I mean.

Delivered

iMessage   Send

DAMN YOU, AUTOCORRECT!

246

Infidelity Party

To: Details

> Hi all, throwing a party for my GtAs tonight from school. After grilling we are opening it up! We will have the infidelity going. I will text when we get the fire going.

> Shoot.. Autocorrect! Firepit, not infidelity! Although, you never know.

Lol

📷 (Text Message) Send

who, what, where?

247

Anyway, it's _definitely_ camping & bbqing out weather, so I figured I'd invite you both down to visit & hang out 😊

12:56 PM

I'm betting that all the cabins are taken on Menstrual Day weekend, but probably not on the others.

One More Beer

who, what, where?

Oprah

GUESS WHOS BLACK!!!

Um... I don't know... Oprah Winfrey?

Hahahaha. I meant GUESS WHO'S BACK! Because I'm back from Austin!

And I'm still white haha

Good to know! Now let's go dwarfing!

Er, DRINKING. This whole convo was a fail.

DAMN YOU, AUTOCORRECT!

250

Getting Steamy

YES OPTUS 2:15 AM 17%

Messages ████████ **Edit**

18/04/2013 5:02 PM

Where u at?

Coles trying to find 3 minutes steamed vagina bags -.-

Vegi!!!!

Vegi bags 😳

FUCKIN LOL

Fml lol

I'm pissin myself

Lol I bet. Stupid auto correct lol

iMessage Send

251

who, what, where?

Fit to Fat

John

the gym first around 5:30

otaaayyy

me wes got a gym membership.. we start today.

Yay :) What gym?

Planet fatness

lol fitness hahaha

wow.. awesome epic fail

who, what, where?

9

Trail Fails

Just can't get
it right

Best Attack

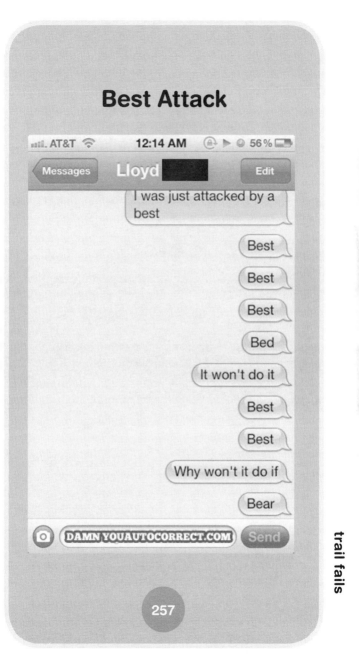

257

trail fails

Dinnertime

5:53PM What did too make for dinner

5:53PM English is my first language. My phone disagrees

5:54PM I'm badgering cocked

5:54PM Seriously this phone is insane

5:55PM I'm making chicken leg quarters, masked parasites

Hehe leftover ham and beans **5:55PM**

5:55PM Mashed pirates

5:55PM Ugh.

5:55PM Makes portraits

5:55PM Mashed potatoes and corn

258

Downward Spiral

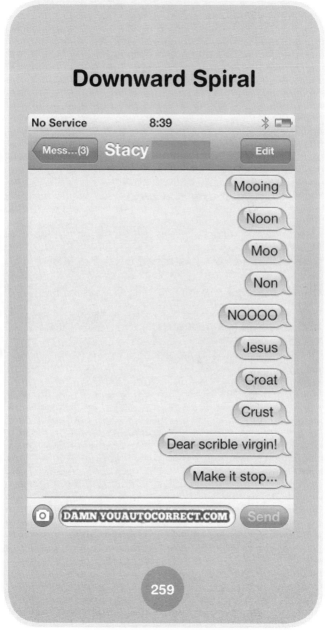

trail fails

Promoted

trail fails

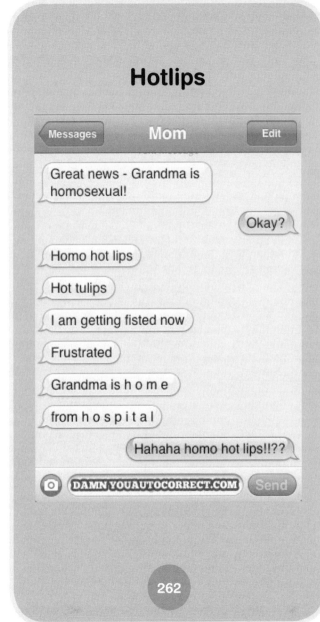

Issues

Oh :(

I know a guy who is allergic to LAWSBLAWWS

Ya :(

duck I ment lobster

Ducking autoturducken

I phone is Sullivan

*stopped

*stupid

Fail -_-

Late crab and and clam

DAMNYOUAUTOCORRECT.COM Send

263

We Need To Talk About Kevin

Messages | **Omar** | **Edit**

No: there have been times that ice taken it off because it was too big and didn't have time to go get it resized

I've Kevin's Christ

Wtf?

Jeeves

Jerbus

Her bus

Are you serious?!!?

I give up

📷 | iMessage | **Send**

DAMN YOU, AUTOCORRECT!

Landlady

trail fails

Pot Comrade

Messages   **Mum**   **Edit**

2 Jun 2012 18:21

> There's isn't a bus for nearly an hour

5 Jun 2012 02:04

Please. Bring my pot comrade

I mean please bring my honey tinkerer

Bloody elves this Phillip is so hat to use

PHOENIX

Never minge

I.gve up

📷 **DAMN YOUAUTOCORRECT.COM**   Send

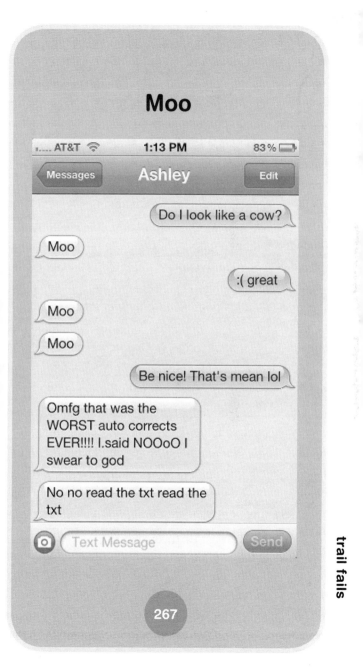

Moo

trail fails

Pancakes

Notifications for this conversation are off. | Turn On

damn auto zucchinis aalways fuckin shit up

Auto cocksucker always corrects my pancakes

Mistakes**
What the FLOUNDER

FISCAL CLIFF*

FOXWORTHY****

NEUROTRANSMITTERS*

Roberto

➕ [   ] Send

DAMN YOU, AUTOCORRECT!

268

Calm Down

trail fails

Pixie

I am sooo hyper right now

Why

Just ate a giant pixie's dick

PIXIE'S DICK

Holy shirtdress

I can't believe that just happened to me

Pixie Stick

Wow you need to slow down!

Send

270

trail fails

Hats and Gigs

Chats **Laiba**
last seen yesterday at 7:47 PM

July 18, 2012

11:06 PM ✓ It was daring hats and gigs

11:07 PM ✓ Raining bats and fogs

11:07 PM ✓ Car and figs

11:07 PM ✓ Oh my good

11:07 PM ✓ God

11:07 PM ✓ Wtf

11:07 PM ✓ Hata n Dodson

11:07 PM ✓ Asdfghjkl

11:08 PM ✓ Cats and dogs

11:08 PM ✓ Yes!!!!!

11:09 PM ✓ Hata n Dodson lol

Send

273

trail fails

Whalecome

274

Darth Evader

Messages   **Jazz (The Wife)**   Edit

Made it to six fags with the dearth vibrator mask

Flags*

Death vibrator

Dearth vader

Farther vader

Dearth

Darth

Hahahahhahahahahhahah ahab

(•_•)

DAMNYOUAUTOCORRECT.COM   Send

275

Skittles

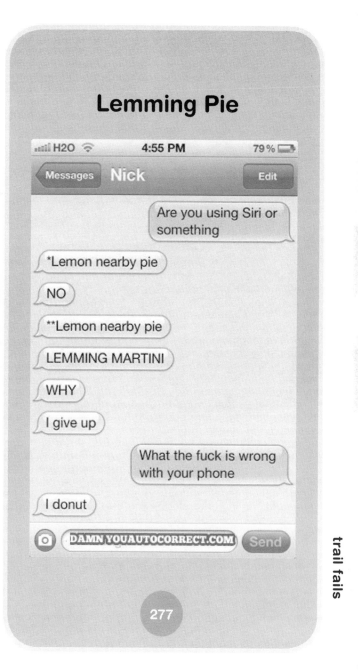

Amazon

Hahahaha. Win.

I had Trix for breakfast! It was so seductive!

*seductive

delicate

delicious

Stupid auto caucus!

auto cucumber

Ughhh never mind, I had Trix this morning and it wa amazon

It was amazing

◎ **DAMN YOUAUTOCORRECT.COM**   Send

trail fails

Dandelions

Hmmm good point. So, did you want the dandelion sex fluids still?

Wed, Apr 17, 2013, 4:32 PM

Seed cliffs.

Wed, Apr 17, 2013, 4:32 PM

Fluffs

Wed, Apr 17, 2013, 4:32 PM

Omg stupid architect

Wed, Apr 17, 2013, 4:33 PM

Autocorrect

Wed, Apr 17, 2013, 4:33 PM

If I ever learn to sing, I'm naming my band dandelion sex fluids.

281

trail fails

Dear Steve Jobs

Messages   Edit

> Dead Steve Jobs, spank your Kindle for the iPoo Ouch

> Sun of a bench

> Dead=deaf
> Spank=Hanks
> Kindle=Kindle
> iPoo=iPoop
> Ouch=pouch
> Your=you're

> On Ho my dog! Holly crust!

> D e a r Steve Jobs, Thank* you* kindly* for the iP o d Touch*

Delivered

Text Message   Send

DAMN YOU, AUTOCORRECT!

282